Copyright © 2017 by Dr. N. J. Raju

All rights reserved.

No part of this book may be reproduced in any written, electronic, recording, or photocopying form without written permission of the author, Dr. N. J. Raju and the publisher. Books may be purchased in quantity and/or special sales by contacting the publisher.

Published by:

CreateSpace Independent Publishing Platform

4900 LaCross Rd.
North Charleston,
South Carolina,
USA-29406

ISBN-13: 978-1547116010

ISBN-10: 1547116013

Preface

This book has been organized in order to provide relevant and practical information on each of the above topics.

We have taught many undergraduate and graduate courses about natural products in plants, i.e., Plant Biotechnology, Practical Botany, and Plants, People, and Environment. As a result, it became apparent there was a need for a comprehensive, thorough collection of information regarding exactly what kinds of natural products plants produce and why they produce them. Currently, such information is contained within thousands of somewhat disjointed reports about the helpful qualities and toxic effects of different plant species throughout the world. The aim of this book is to help bring more unity and understanding to this complicated and often contradictory jumble of information. Such a book is a necessity for many, including: biochemists, natural product chemists, and molecular biologists; research investigators in industry, federal labs, and universities; physicians, nurses, and nurse practitioners; pre-medical and medical students; ethnobotanists, ecologists, and conservationists; nutritionists; organic gardeners and farmers; and those interested in herbs and herbal medicine. With the growing interest in this field by professionals and the general public alike, it was important for us to produce a book that encompasses as much information as possible on the natural products produced by plants as well as their importance in today's world.

Acknowledgments

We thank the following individuals who have assisted us or provided us with information in the preparation of this book.

- Students at the Andhra University who worked on the case studies with individual plants and the important natural products that they produce.
- Dr. Lakshmi Prasanna, who provided us with information on the Book editing.

CONTENTS

S.NO	TITLE	PAGE NO.
1.	Introduction to Inflammation	1-7
2.	Inflammatory disorders	8
3.	Plants with antiinflammation activity	9-11
4.	Evaluation of anti-inflammatory activity	12-23
5.	Effect of ethyl acetate extract of roots and rhizomes of *Antigonon leptopus*	24-26
6.	Effect of methanol extract of roots and rhizomes of *Antigonon leptopus*	27-29
7.	Results and Discussion	30
8.	Effect of Ethyl acetate extract of roots of *Glycosmis pentaphylla*	31-33
9.	Effect of Methanol extract of roots of *Glycosmis pentaphylla*	34-36
10.	Results and Discussion	37-38
11.	Effect of Ethyl acetate extract of stem heart wood of *Spondias pinnata*	39-41
12.	Effect of Methanol extract of roots of *Spondias pinnata*	42-44
13.	Results and Discussion	45-46
14.	Conclusion Of Anti-Inflammatory Activity	47-48
15.	References	49-52

1. Introduction:

Inflammation (Latin, *inflammatio*, to set on fire) is the complex biological response of vascular tissues to harmful stimuli, such as pathogens, damaged cells, or irritants. It is a protective attempt by the organism to remove the injurious stimuli as well as initiate the healing process for the tissue. Inflammation is not a synonym for infection. Even in cases where inflammation is caused by infection it is incorrect to use the terms as synonyms: infection is caused by an exogenous pathogen, while inflammation is the response of the organism to the pathogen.

In the absence of inflammation, wounds and infections would never heal and progressive destruction of the tissue would compromise the survival of the organism. However, inflammation which runs unchecked can also lead to a host of diseases, such as hay fever, atherosclerosis, and rheumatoid arthritis. It is for this reason that inflammation is normally tightly regulated by the body.

Inflammation can be classified as either *acute* or *chronic*. *Acute inflammation* is the initial response of the body to harmful stimuli and is achieved by the increased movement of plasma and leukocytes from the blood into the injured tissues. A cascade of biochemical events propagates and matures the inflammatory response, involving the local vascular system, the immune system, and various cells within the injured tissue. Shown in table-1.1 & 1.2. Prolonged inflammation, known as *chronic inflammation*, leads to a progressive shift in the type of cells which are present at the site of inflammation and is characterised by simultaneous destruction and healing of the tissue from the inflammatory process [1].

Table-1.1: The classic signs and symptoms of acute inflammation: [1]

English	Latin
Redness	*Rubor*
Heat	*Calor*
Swelling	*Tumor*
Pain	*Dolor*
Loss of function	*Functio laesa*

Chronic inflammation is a pathological condition characterised by concurrent active inflammation, tissue destruction, and attempts at repair. Chronic inflammation is not characterised by the classic signs of acute inflammation listed above. Instead, chronically inflamed tissue is characterised by the infiltration of mononuclear immune cells (monocytes, macrophages, lymphocytes, and plasma cells), tissue destruction, and attempts at healing, which include angiogenesis and fibrosis[2].

Endogenous causes include persistent acute inflammation. Exogenous causes are varied and include bacterial infection, especially by *Mycobacterium tuberculosis*, prolonged exposure to chemical agents such as silica, tobacco smoke, or autoimmune reactions such as rheumatoid arthritis.

Table-1.2.: Comparison between acute and chronic inflammation: [1]

	Acute	**Chronic**
Causative agent	Pathogens, injured tissues	Persistent acute inflammation due to non-degradable pathogens,

		persistent foreign bodies, or autoimmune reactions
Major cells involved	Neutrophils, mononuclear cells (monocytes, macrophages)	Mononuclear cells (monocytes, macrophages, lymphocytes, plasma cells), fibroblasts
Primary mediators	Vasoactive amines, eicosanoids	IFN-γ and other cytokines, growth factors, reactive oxygen species, hydrolytic enzymes
Onset	Immediate	Delayed
Duration	Few days	Up to many months, or years
Outcomes	Healing, abscess formation, chronic inflammation	Tissue destruction, fibrosis

1.2.1 Mediators are classified into two types:

Table-1.2.3: a) Plasma derived mediators: [3]

Name	Produced by	Description
Bradykinin	*Kinin system*	A vasoactive protein which is able to induce vasodilation, increase vascular permeability, cause smooth muscle contraction, and induce pain.
C3	*Complement system*	Cleaves to produce *C3a* and *C3b*. C3a stimulates histamine release by mast cells, thereby producing vasodilation. C3b is able to bind to bacterial cell walls and act as an opsonin, which marks the invader as a target for phagocytosis.
C5a	*Complement system*	Stimulates histamine release by mast cells, thereby producing vasodilation. It is also able to act as a chemoattractant to direct cells via chemotaxis to the site of inflammation.

Factor XII (*Hageman Factor*)	*Liver*	A protein which circulates inactively, until activated by collagen, platelets, or exposed basement membranes via conformational change. When activated, it in turn is able to activate three plasma systems involved in inflammation: the kinin system, fibrinolysis system, and coagulation system.
Membrane attack complex	*Complement system*	A complex of the complement proteins C5b, C6, C7, C8, and multiple units of C9. The combination and activation of this range of complement proteins forms the *membrane attack complex*, which is able to insert into bacterial cell walls and causes cell lysis with ensuing death.
Plasmin	*Fibrinolysis system*	Able to break down fibrin clots, cleave complement protein C3, and activate Factor XII.
Thrombin	*Coagulation system*	Cleaves the soluble plasma protein fibrinogen to produce insoluble fibrin, which aggregates to form a blood clot. Thrombin can also bind to cells via the PAR1 receptor to trigger several other

		inflammatory responses, such as production of chemokines and nitric oxide.

Table-1.2.4: b) Cell derived mediators: [4] [5]

Lysosome granules
Histamine
IFN-γ
IL-8
Leukotriene B4
Nitric oxide
Prostaglandins
TNF-α and IL-1

1.2.2 Inflammatory disorders:

Abnormalities associated with inflammation comprise a large, unrelated group of disorders which underly a variety of human diseases. The immune system is often involved with inflammatory disorders, demonstrated in both allergic reactions and some myopathies, with many immune system disorders resulting in abnormal inflammation. Non-immune diseases with aetiological origins in inflammatory processes are thought to include cancer, atherosclerosis, and ischaemic heart disease [3]. Examples of disorders associated with inflammation include:

- Asthma
- Autoimmune diseases
- Chronic inflammation
- Chronic prostatitis
- Glomerulonephritis
- Hypersensitivities
- Inflammatory bowel diseases
- Pelvic inflammatory disease
- Reperfusion injury
- Rheumatoid arthritis
- Transplant rejection
- Vasculitis

1.2.3 Plants with antiinflammation activity:

Inflammation disease including different types of rheumatic diseases is very common throughout the world. Although rheumatism is one of the oldest known diseases of mankind and affects a large number of populations of the world, no permanent cure has been found so far [6].

The greatest disadvantage of the presently available potent synthetic drugs lies in their toxic symptoms in gastric intestinal tract and discontinuation of medication after the treatment. Some plants having anti-inflammatory activity shown in table-1.2.5.

Table-1.2.5: Some plant sources having anti inflammatory activity: [7,8]

Plant name	Trade name in India	Family
Aconitum hapellus	Aconite	Ranunculaceaea
Alpinia officinarum	Rasna	Zingiberaceae
Azadirachta indica	Neem	Meliaceae

Balanites roxburghii	Gari	Simarubiaceae
Boerhaavia diffusa	Punarnava	Nyctanginaceae
Colchicum autumnale	Colchicum	Liliaceae
Curcuma longa	Turmeric	Zingiberaceae
Delonix elata	Vatanarayana	Leguminosae
Glycerrhiza glabra	Liquorice	Leguminosae
Hedychium spicatum	Karpura kacheri	Zingiberaceae
Hellitropium curassavicum	Haatisura	Boraginaceae
Hemidesmus indicus	Margabi	Asclepiadance
Hibiscus rosasinensis	Jassoon	Malvaceae
Indigofera aspalathoides	Hakna	Compositaceae
Innula kashmiriana	Padmapushkara	Iridaceae
Lawsonia innermis	Henna	Lythraceae
Leucas aspera	Hulkusha	Labiatae
Moringa oleifera	Sahinjan	Moringaceae
Morus alba	Tutri	Moraceae
Nerium indicum	Kaner	Apocynaceae
Nyphhaea stellata	Nikamal	Nymphaeaceae
Operulina turpethum	Nakpatra	Convolvulaceae
Ougenia oojeinesis	Sandan	Febaceae
Paederia foetida	Gandhali	Rubiaceae
Phyla nodiflora	Jalapeepala	Verbenaceae

Piper longum	Pippali	Piperaceae
Randia dumetorum	Mainphal	Rubiaceae
Salvadora percica	Brihatpilu	Salvadoraceae
Teramnus labialis	Mashaparni	Leguminosae
Tinospora malabarica	Sudarsana	Menispermaceae
Urena iobata	Vanabhendra	Malveceae
Vanda roxburghii	Rasna	Orchidaceae
Verbena officinalis	Karaita	Verbenaceae
Vitex negundo	Nirgundi	Verbenaceae
Withania somnifera	Ashwagandha	Slanaceae

The literature survey reveals that so many medicinal plants species exhibits anti-inflammatory activity. Some of the plant sources are used in traditional systems of medicines with pharmacological and therapeutically proven anti-inflammatory and anti rheumatic claims [9].

1.2.4 Evaluation of anti-inflammatory activity:

The complexity of inflammatory process and the diversities of the drugs that have been found effective in modifying the process have resulted in the development of numerous methods for detecting anti-inflammatory substances. In the present investigation, the author has tried to test the anti inflammatory activity of ethyl acetate and methanolic extract of the plants *Antigonon leptopus* (roots and rhizomes), *Glycosmis pentaphylla* (roots) and *Spondias pinnata* (stem heart wood). The method that followed was carrageenan induced rat paw edema model [10].

1.2.5 Acute models of inflammation [11,12]

a) **Carrageenan induced oedema model**

Acute hind paw oedema was induced either in mice or in rats by injecting 0.05 ml to 0.1 of 1 % w/v carrageenan which reaches a peak level at 3-5 hrs of carrageenan injection. Although oedema can be induced by many other phlogistic agents like dextrin, formaldehyde, 5-hydroxytryptamine, histamine bradykinin and prostaglandin E1 etc., for routine screening, acute carrageenan induced oedema test was employed.

b) **U.V.light induced erythemea model**

Exposure to U.V. radiation also induces acute erythemea which is used as model for anti-inflammatory activity testing.

1.2.6 Chronic models of inflammation
a) Cotton pellet test [13,14]

Chronic inflammation was induced by the implanatation of sterile cotton pellets (50 mg ± 1 mg) on the back or axilla of the rats asceptically. The peak effect was reached within 7 days.

b) Granuloma pouch test [15,16]

Pouch on the back of the rat was produced by injecting 20 ml of air and 1.0 ml of 1% croton oil in olive oil or 0.5ml of turpentine oil in the subcutaneous tissues in between the shoulder blades. The effect was seen after 7 days.

c) Formaldehyde induced arthritis [17,18]

Arthritis was induced by injecting 0.1 ml of 2 % formaldehyde solution into the sub plantar region of one of the hind paws of rat on the first and third day of the 10 days experiment.

d) Adjuvant induced arthritis [19,20]

Chronic arthritis in rats was induced by injection of 0.5 mg of killed *Mycobacterium tuberculosis* (Difco) suspended in 0.1 ml of liquid paraffin into one of the hind paws. The effect was observed till 40 days of irritant injection.

1.2.7 Method used in the present study [Carrageenan-induced rat paw oedema model for assessment of acute inflammation]: [21]

Subcutaneous carrageenan injection to produce oedema in the rat paw is the most frequently used acute inflammatory animal model among the other models such as UV light induced erythemea model etc. Various measuring systems used for the assessment of induced oedema in the paw include: volume [22]; paw thickness [23]; paw weight [24]; and painfulness [25] to monitor the development of the induced oedema in the paw. The preferred routine system in our laboratory is measurement of dorsiventral paw thickness using zeitlin's apparatus (unpublished) (Fig-3.2.1).

1.2.8 Apparatus available for measurement of oedema (paw thickness/volume):

1. zeitlin's constant loaded lever[26]
2. plethysmograph[27]

1.2.9 Evaluation of Model

To evaluate this model, the percentage increase in paw thickness was plotted against the time (Hour) and the maximal oedema response induced during the 6 hours was determined. The results showed the ability of the model in detecting that the time course changes in the paw size was associated with carrageenan induced rat paw oedema. The paw oedema was constantly increased during 4 hours and reached peak of oedema at 4th

hour. At the 5th and 6th hour, the oedema was gradually reduced.

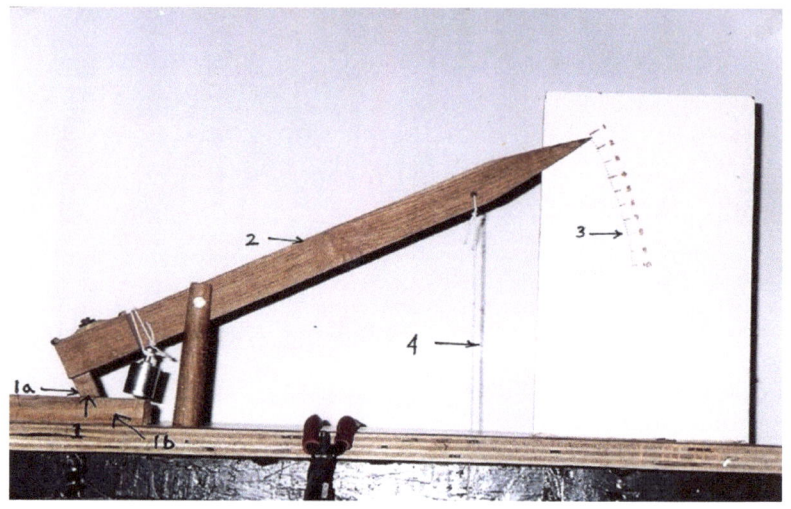

Zeitlin's Apparatus

(Fig-1.2.1)

1. Place where the paws are to be kept to measure the thickness.
2. Constant load lever.
3. Graduates scale number between 1-10 and divided by 0.5 equal to 20 divisions.
4. Thread to pull down the lever with right leg in order to facilitate to keep the paw in between pointer (1a) and basement (1b).

Progression of the carrageenan-induced rat paw oedema over 6h as monitored with Zeitlin's apparatus

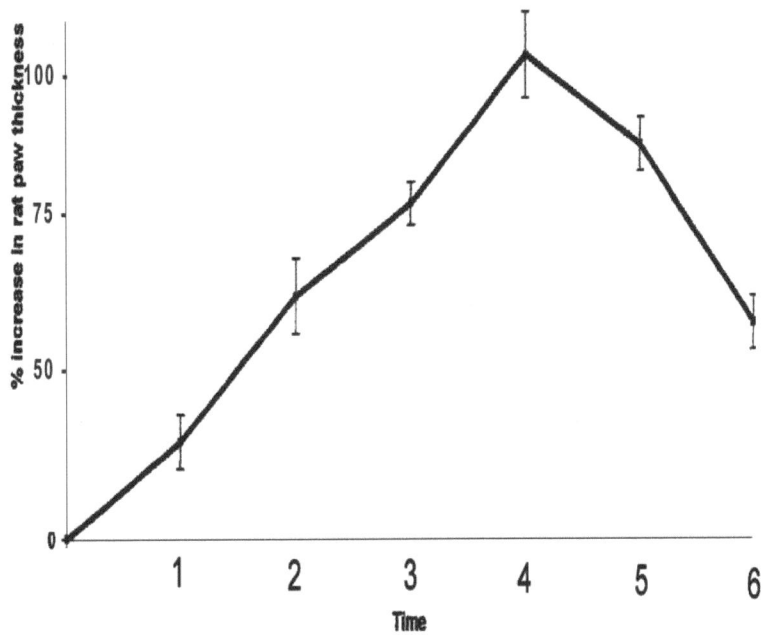

3.2.10 Materials and methods:

Animals: Albino wistar rats of either sex (175-250gm)
Carrageenan   : 1% suspension in saline
Standard drug   : Ibuprofen (2.68mg/kg)
Drug vehicle   : Sodium CMC in water

A) Materials:

All the materials used for this experiment are of Pharmacopoeial grade. Ibuprofen were purchased from local supplier (Sai chemicals) manufactured by Ranbaxy Diagnostics Ltd., New Delhi, India. Water represents the double distilled water, standard orogastric cannula was used for oral drug administration.

B) Animals:

Wistar albino rats of either sex weighing between 175-250 gm were obtained from M/s. Mahavir Enterprises, Hyderabad, Andhra Pradesh, India. The animals were housed under standard environmental conditions (temperature of 22 ± 1°C with an alternating 12 h light–dark cycle and relative humidity of 60 ± 5 %), one week before the start and also during the experiment as per the rules and regulations of the Institutional Animal Ethics committee and by the Regulatory body of the government (Regd no. 516/01/a/CPCSEA). They were fed with standard laboratory diet supplied by M/s. Rayans biotechnologies Pvt. Ltd., Hyderabad,

Andhra Pradesh, India. Food and water was allowed *ad libitum* during the experiment.

C) Preparation of carrageenan suspension:

Suspension of carrageenan sodium salt 1% was prepared by sprinkling 100mg of carrageenan powder in 10 ml of saline (0.9%NaCl) and set a side to soak for 1 hr, and then the suspension was mixed thoroughly.

D) Preparation of sodium CMC suspension:

Stock suspension of sodium CMC was prepared by triturating the powder sodium CMC (1g) finely in 2.5 ml of water containing tween 20. A 1:10 dilution of this stock solution made in distilled water was used for suspending the test and standard drugs.

1.2.11 Measurement of oedema:

All these treatments were given orally according to their body weight 1 hour prior to the induction of oedema.

Before induction of the edema, the dorsiventral thickness of both the paws of each was measured using Zetlin's apparatus, which consists of a graduated micrometer, combined loaded lever system to magnify the small changes in paw thickness during the course experimental and also measurements were measured by Vernier calipers. The measurements were taken at 1 hr intervals after induction of oedema for up to 3 hrs. Oedema was monitored as the percentage increase in paw thickness in the carrageenan increase in paw thickness produced in the saline injected paw was subtracted from that of carrageenan injected left paw.

Percentage increase in paw thickness $(Y_t - Y_0)/Y_0 \times 100$

Y_t = paw thickness at the time 't' hours (after injection)

Y_0 = paw thickness at the time '0' hours

The percentage increase in paw thickness during 3 hours was determined. The percent inhibition of paw oedema thickness is calculated using the formula

Percentage inhibition = $100[1 - Y_t/Y_c]$

Y_t = average increase in paw thickness in groups tested with test compounds

Y_c = average increase in paw thickness in control

1.2.12 Experimental Protocol

The rats were pre dosed orally with extracts at different dose levels 18 hours and 2 hours (unless otherwise mentioned) prior to the induction of carrageenan subcutaneously (SC) into the sub plantar tissue of the hind paw of each rat, 0.1 ml of 1% carrageenan suspension.

The drug effects were estimated by comparing the maximal oedema response during 6 hours in the drug as extract treated group with that of vehicle treated group as control. Group 1- normal rats treated with vehicle (1% Sodium CMC) and served as normal control and Group 2- Ibuprofen 50 mg/kg b.w, Group 3-5 were treated with the ethyl acetate extract of **Antigonon leptopus**(roots and rhizomes), Group 6-8 were treated with the methanol extract of **Antigonon leptopus**(roots and rhizomes), at doses of 100, 200 and 400mg/kg b.w. respectively, Group 9-11 rats were treated with ethyl acetate extract of **Glycosmis pentaphylla**(roots) Group 12-14 were treated with methanol extract of **Glycosmis pentaphylla**(roots) at doses of 100, 200 and 400 mg/kg b.w, respectively, Group 15-17 were treated with ethyl acetate extract of **Spondias pinnata**(stem heart wood), at doses of 100, 200, 400 mg/kg b.w., respectively and Group 18-20 were treated with methanol extract of *Spondias pinnata*(stem heart wood), at doses of 100, 200, 400 mg/kg b.w., respectively. All the doses were administered orally according to the body weight of the animals.

1.2.13 Statistical analysis:

The results are expressed as mean ± s.e.m and the statistical significance of difference between groups was analyzed by "t" test (two tailed) followed by Dunnett's multiple comparison test $p<0.05$ was considered as significant.

Calculations were done using Prism pad software version-5.

1.2.14 Drug effects:

The rats were always pre-dosed orally with the extracts or purified compounds 18 hours and 2 hours (unless otherwise mentioned) prior to the induction of carrageenan.

The drug effects were estimated by comparing the maximal oedema response during 6 hours (monitored as % increase in paw thickness) in the drug or extract treated group with that of drug vehicle treated group as control.

1.2.15 Results:

Carrageenan has produced significant oedema in the left hind paw of the vehicle treated group and the paw oedema was significantly reduced (P<0.001) in the standard drug, Ibuprofen (50 mg/kg) treated group.

Ethyl acetate extract of **Antigonon leptopus**(roots and rhizomes), **Glycosmis pentaphylla**(roots) and **Spondias pinnata**(stem heart wood), at doses of 100, 200 and 400mg/kg; p.o respectively, extracts at the dose of 400 mg/kg exhibited significant reduction (P<0.001) in paw thickness when compared to control group treated with standard drug Ibuprofen (50 mg/kg) at third and fourth hour. The results were tabulated:

1.2.16 Effect of ethyl acetate extract of roots and rhizomes of *Antigonon leptopus*:

The animals were divided into five groups of four animals each. Roots and rhizomes of *Antigonon leptopus* ethyl acetate extracts produced significant reduction in paw thickness at all the treated doses (100, 200 and 400 mg/kg) which are comparable with that of standard drug ibuprofen (50mg/kg). The effect produced by the extracts is dose dependent. The results were given in Fig-1.2.2 and Table-1.2.6.

Table–1.2.6: Percentage inhibition of carrageenan induced paw oedema in rats:

S. no.	Treatment	% inhibition maximal Paw oedema	% inhibition of total (AUC) paw oedema
1.	Drug vehicle	0.0 ± 1.95	0.0 ± 7.47

2.	Ibuprofen (50 mg.kg^{-1})	72.41 ± 1.49**	79.65 ± 2.29**
3.	AL (100 mg. kg^{-1})	61.22 ± 1.78*	60.42 ± 1.06*
4.	AL (200 mg. kg^{-1})	68.06 ± 0.80**	68.50 ± 0.45**
5.	AL (400 mg. kg^{-1})	66.34 ± 2.15**	71.28 ± 1.26**

P: *<0.05, **<0.01, ***<0.001, ns=not significant

A)

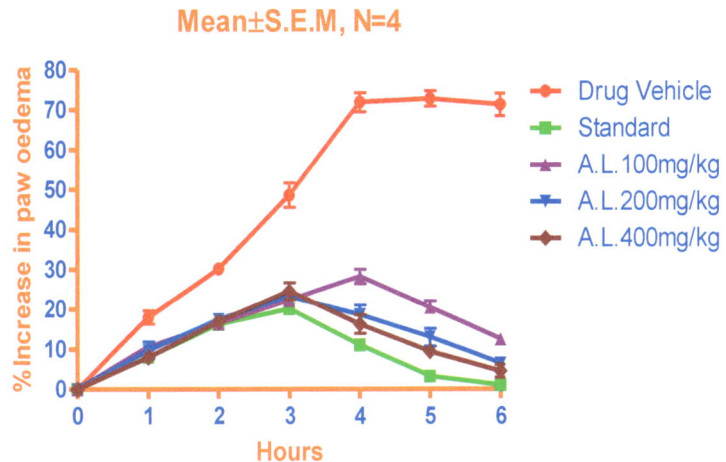

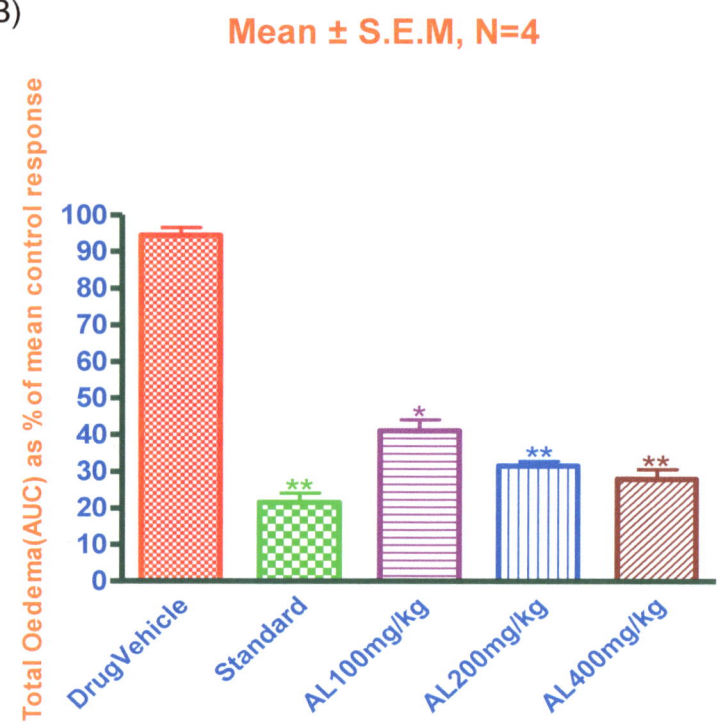

Fig-1.2.2: Effects of the crude ethyl acetate extracts of *Antigonon leptopus*(roots and rhizomes),100, 200 and 400 mg/kg respectively along with Ibuprofen 50 mg/kg on A) the maximal and B) the total paw oedema in carrageenan induced rats.

1.2.17 Effect of methanol extract of roots and rhizomes of *Antigonon leptopus*:

The animals were divided into five groups of four animals each. Roots and rhizomes of *Antigonon leptopus* methanol extracts produced significant reduction in paw thickness at all the treated doses (100, 200 and 400 mg/kg) which are comparable with that of standard drug ibuprofen (50mg/kg). The effect produced by the extracts is dose dependent. The results were given in Fig-1.2.3 and Table-1.2.7.

Table–1.2.7: Percentage inhibition of carrageenan induced paw oedema in rats:

S. no.	Treatment	% inhibition maximal paw oedema	% Inhibition of total (AUC) paw oedema
1.	Drug vehicle	0.0 ± 1.95	0.0 ± 7.47
2.	Ibuprofen(50 mg.kg^{-1})	72.41 ± 1.49**	79.65 ± 2.29**

6.	AL (100 mg. kg^{-1})	64.62 ± 1.26*	64.67 ± 2.44*
7.	AL (200 mg. kg^{-1})	69.60 ± 1.30**	75.41 ± 0.46**
8.	AL (400 mg. kg^{-1})	68.04 ± 0.80**	72.65 ± 1.44**

P: *<0.05, **<0.01, ***<0.001, ns=not significant

A)

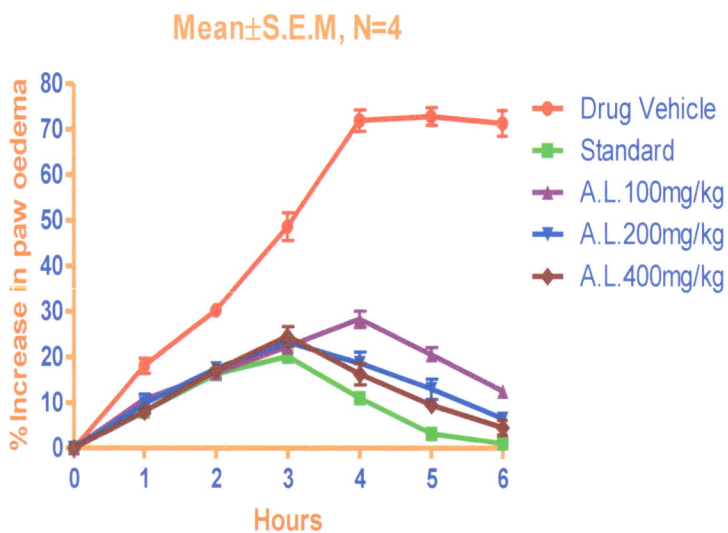

B)

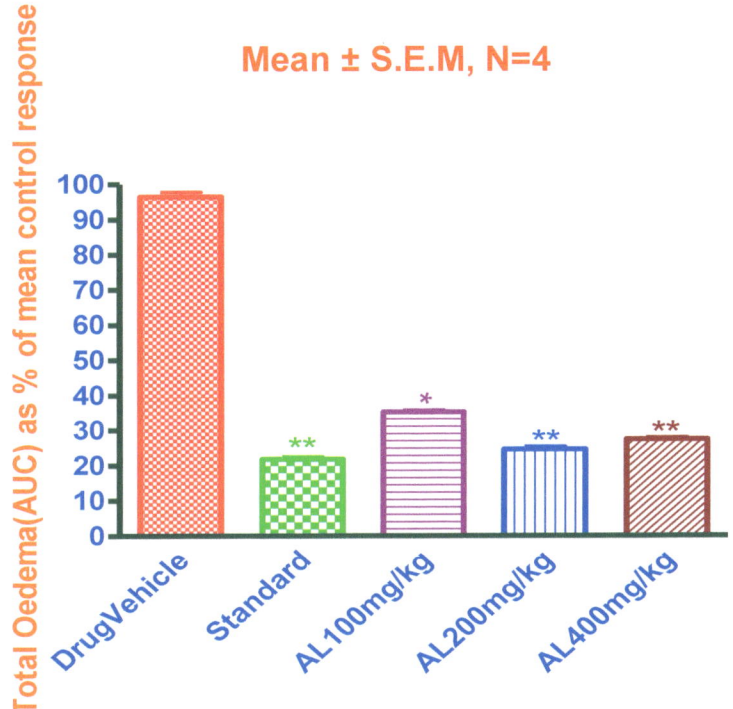

Fig-1.2.3: Effects of the crude methanol extracts of *Antigonon leptopus*(roots and rhizomes), 100, 200 and 400 mg/kg respectively along with Ibuprofen 50 mg/kg on A) the maximal and B) the total paw oedema in carrageenan induced rats.

Results and Discussion:

There was significant and dose dependent anti-inflammatory activity of both the ethyl acetate and methanol extracts in the acute Carrageenan-induced rat paw oedema model. Orally administered doses of 100, 200 and 400mg/kg of both ethyl acetate (61.22 ± 1.78, 68.06 ± 0.80, 66.34 ± 2.15) and methanol (64.62 ± 1.26, 69.60 ± 1.30, 68.04 ± 0.80) extracts of roots and rhizomes of *A. leptopus* produced significant reduction in paw oedema, as compared to Ibuprofen (72.41 ± 1.49) (standard) 50mg/kg. Doses of 200mg/kg of both the ethyl acetate and methanol extracts showed significant activity.

Prostaglandins and bradykinins were suggested to play an important role in Carrageenan-induced oedema [28,29]. As phytochemical tests showed the presence of sterols, flavonoids and tannins in both ethyl acetate and methanol extracts they might suppress the formation of prostaglandins and bradykinins or antagonize their action and exert its activity [30]. Further studies are needed to explore mechanism by which *A. leptopus* produces anti-inflammatory activity.

1.2.18 Effect of Ethyl acetate extract of roots of *Glycosmis pentaphylla*:

The animals were divided into five groups of four animals each. Roots of Glycosmis *pentaphylla* ethyl acetate extracts produced significant reduction in paw thickness at all the treated doses (100, 200 and 400 mg/kg) which are comparable with that of standard drug ibuprofen (50mg/kg). The effect produced by the extracts is dose dependent. The results were given in Fig.1.2.4 and Table-1.2.8.

Table–1.2.8: Percentage inhibition of carrageenan induced paw oedema in rats:

S. no.	Treatment	% inhibition maximal Paw oedema	% inhibition of total (AUC) paw oedema
1.	Drug vehicle	0.0 ± 1.95	0.0 ± 7.47
2.	Ibuprofen(50 mg.kg^{-1})	72.41 ± 1.49**	79.65 ± 2.29**
9.	GP (100 mg. kg-1)	62.59 ± 1.76*	64.57 ± 0.74*

| 10. | GP (200 mg. kg-1) | 70.26 ± 0.52** | 74.02 ± 0.98** |
| 11. | GP (400 mg. kg-1) | 71.73 ± 0.89** | 74.61 ± 1.62** |

P: *<0.05, **<0.01, ***<0.001, ns=not significant

A)

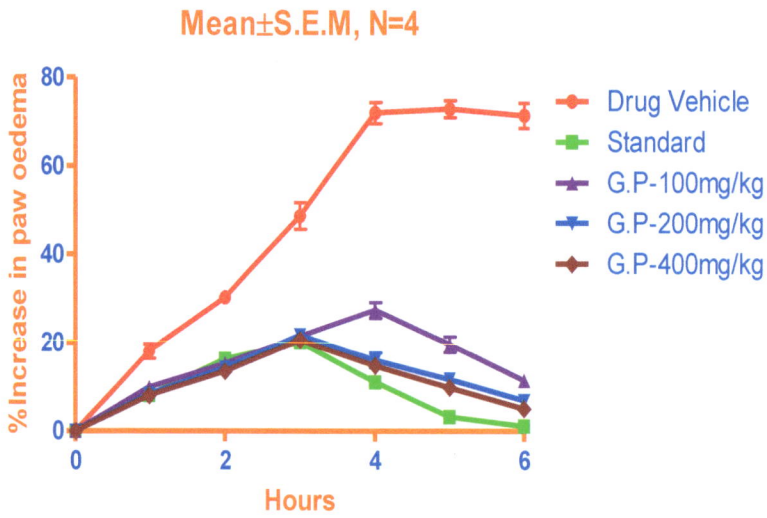

B)

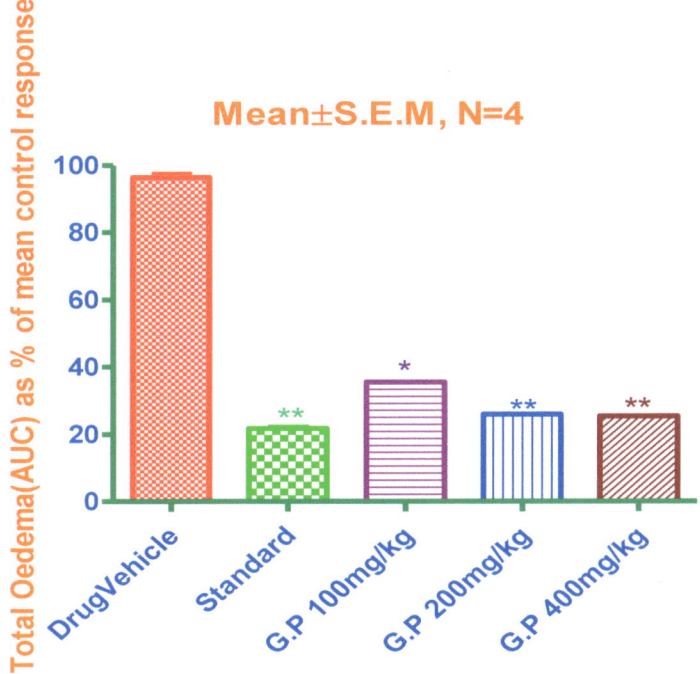

Fig-1.2.4: Effects of the crude ethyl acetate extracts of *Glycosmis pentaphylla* (roots) 100, 200 and 400 mg/kg respectively along with Ibuprofen 50 mg/kg on A) the maximal and B) the total paw oedema in carrageenan induced rats.

3.2.19 Effect of Methanol extract of roots of *Glycosmis pentaphylla*:

The animals were divided into five groups of four animals each. Roots of Glycosmis *pentaphylla* methanol extracts produced significant reduction in paw thickness at all the treated doses (100, 200 and 400 mg/kg) which are comparable with that of standard drug ibuprofen (50mg/kg). The effect produced by the extracts is dose dependent. The results were given in Fig.1.2.5 and Table-1.2.9.

Table–1.2.9: Percentage inhibition of carrageenan induced paw oedema in rats:

S. no.	Treatment	% inhibition maximal paw oedema	% Inhibition of total (AUC) paw oedema
1.	Drug vehicle	0.0 ± 1.95	0.0 ± 7.47
2.	Ibuprofen(50 mg.kg^{-1})	72.41 ± 1.49**	79.65 ± 2.29**

12.	GP (100 mg. kg^{-1})	62.21 ± 1.65*	64.09± 1.62*
13.	GP (200 mg. kg^{-1})	67.23 ± 0.57**	71.75 ± 0.60**
14.	GP (400 mg. kg^{-1})	69.96 ± 0.74**	73.62 ± 1.00**

P: *<0.05, **<0.01, ***<0.001, ns=not significant

A)

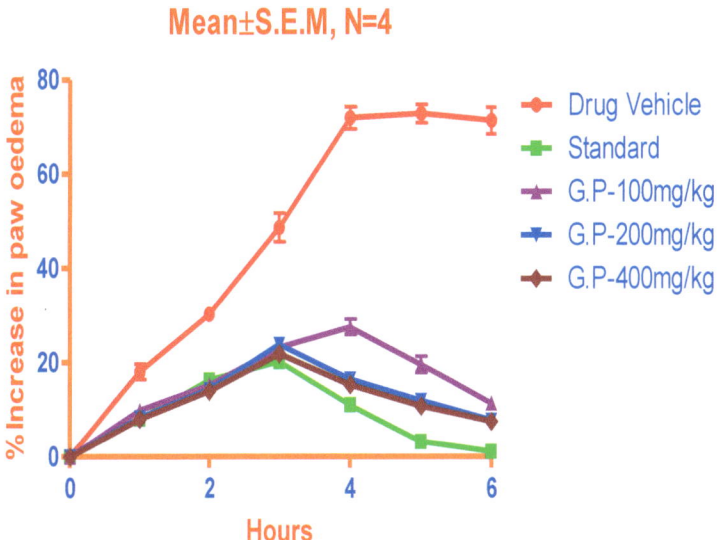

B)

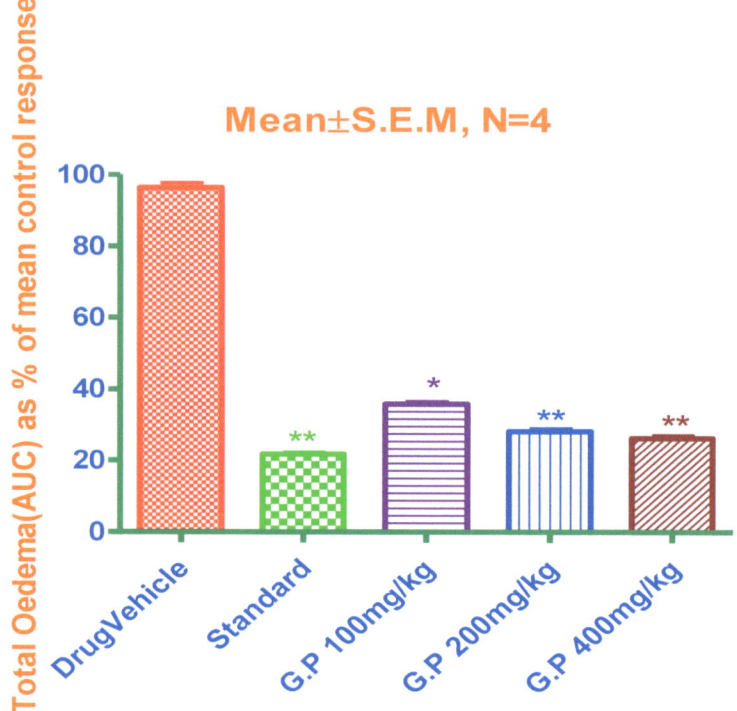

Fig-1.2.5: Effects of the crude methanol extracts of *Glycosmis pentaphylla* (roots), **100, 200 and 400 mg/kg** respectively along with Ibuprofen 50 mg/kg on A) the maximal and B) the total paw oedema in carrageenan induced rats.

Results and Discussion:

There was significant and dose dependent anti-inflammatory activity of both the ethyl acetate and methanol extracts in the acute Carrageenan-induced rat paw oedema model. Orally administered doses of 100, 200 and 400mg/kg of both ethyl acetate (62.59 ± 1.76, 70.26 ± 0.52, 71.73 ± 0.89) and methanol (62.21 ± 1.65, 67.23 ± 0.57, 69.96 ± 0.74) extracts of roots and rhizomes of *G. pentaphylla* produced significant reduction in paw oedema, as compared to Ibuprofen (standard)(72.41 ± 1.49) 50mg/kg.

Carrageenan has been widely used as a noxious agent able to induce experimental inflammation for the screening of compounds possessing anti-inflammatory activity [32]. This phlogistic agent, when injected locally into the rat paw, produced a severe inflammatory action, which was discernible within 30 min [33]. The development of oedema induced by carrageenan corresponds to the events in the acute phase of inflammation, mediated by histamine, Prostaglandins and bradykinins produced under an effect of cycloxygenase [34]. As phytochemical tests showed the presence of glycosides, flavonoids and alkaloids in both ethyl acetate and methanol extracts they might suppress the formation of prostaglandins and bradykinins or antagonize their action and exert its activity [30]. Further studies are needed to explore mechanism by which *G. pentaphylla* produces anti-inflammatory activity.

1.2.20 Effect of Ethyl acetate extract of stem heart wood of *Spondias pinnata*:

The animals were divided into five groups of four animals each. Stem heart wood of *Spondias pinnata* ethyl acetate extracts produced significant reduction in paw thickness at all the treated doses (100, 200 and 400 mg/kg) which are comparable with that of standard drug ibuprofen (50mg/kg). The effect produced by the extracts is dose dependent. The results were given in Fig-1.2.6 and Table-1.2.10.

Table–1.2.10: Percentage inhibition of carrageenan induced paw oedema in rats:

S. no.	Treatment	% inhibition of maximal Paw oedema	% inhibition of total (AUC) paw oedema
1.	Drug vehicle	0.0 ± 1.95	0.0 ± 7.47
2.	Ibuprofen(50 mg.kg^{-1})	72.41 ± 1.49**	79.65 ± 2.29**
15.	SP (100 mg. kg^{-1})	64.32 ± 0.74*	64.81 ± 0.64*

| 16. | SP (200 mg. kg⁻¹) | 69.04 ± 0.29** | 73.04 ± 0.46** |
| 17. | SP (400 mg. kg⁻¹) | 72.10 ± 0.45** | 75.20 ± 0.58** |

P: *<0.05, **<0.01, ***<0.001, ns=not significant

A)

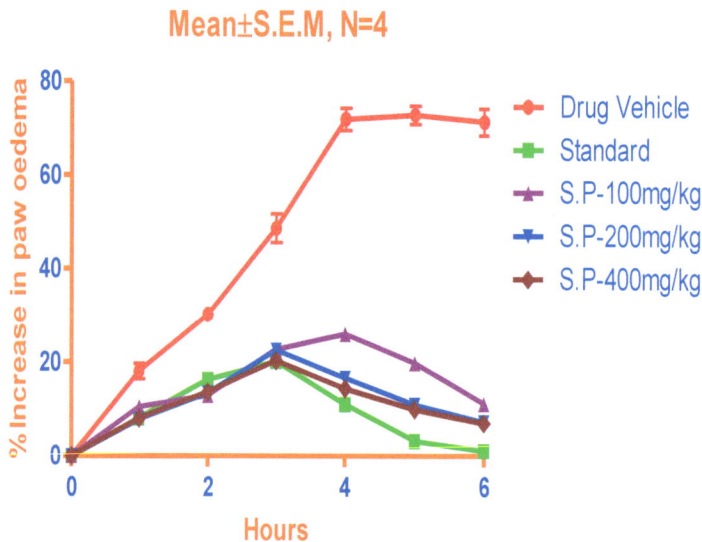

B)

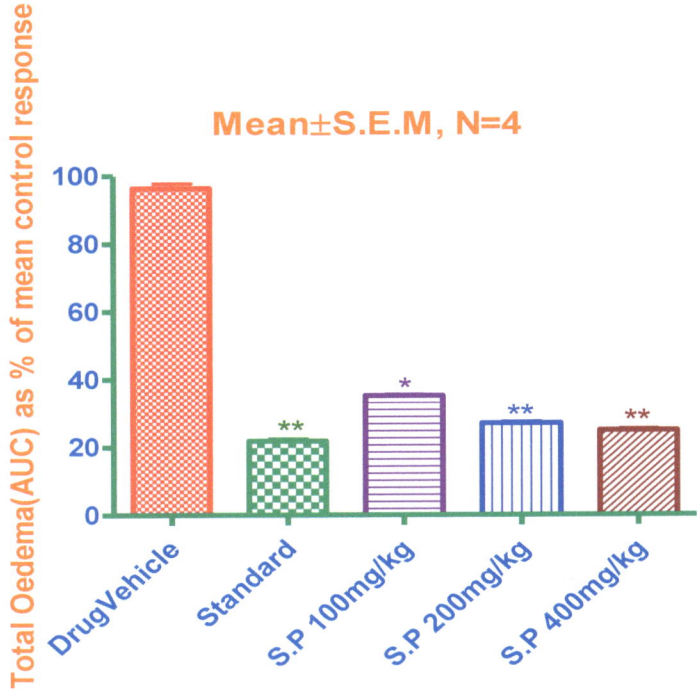

Fig-1.2.6: Effects of the crude ethyl acetate extracts of *Spondias pinnata* (stem heart wood), 100, 200 and 400 mg/kg respectively along with Ibuprofen 50 mg/kg on A) the maximal and B) the total paw oedema in carrageenan induced rats.

1.2.21 Effect of Methanol extract of roots of *Spondias pinnata*:

The animals were divided into five groups of four animals each. Stem heart wood of *Spondias pinnata* methanol extracts produced significant reduction in paw thickness at all the treated doses (100, 200 and 400 mg/kg) which are comparable with that of standard drug ibuprofen (50mg/kg). The effect produced by the extracts is dose dependent. The results were given in Fig-1.2.8 and Table-1.2.11.

Table–1.2.11: Percentage inhibition of carrageenan induced paw oedema in rats:

S. no.	Treatment	% inhibition maximal paw oedema	% Inhibition of total (AUC) paw oedema
1.	Drug vehicle	0.0 ± 1.95	0.0 ± 7.47
2.	Ibuprofen(50mg.kg^{-1})	72.41 ± 1.49**	79.65 ± 2.29**

18.	SP (100 mg. kg^{-1})	63.67 ± 1.26*	65.30 ± 0.54*
19.	SP (200 mg. Kg^{-1})	68.82 ± 0.32**	72.85 ± 0.35**
20.	SP (400 mg. kg^{-1})	71.71 ± 0.68**	75.30 ± 0.50**

P: *<0.05, **<0.01, ***<0.001, ns=not significant

A)

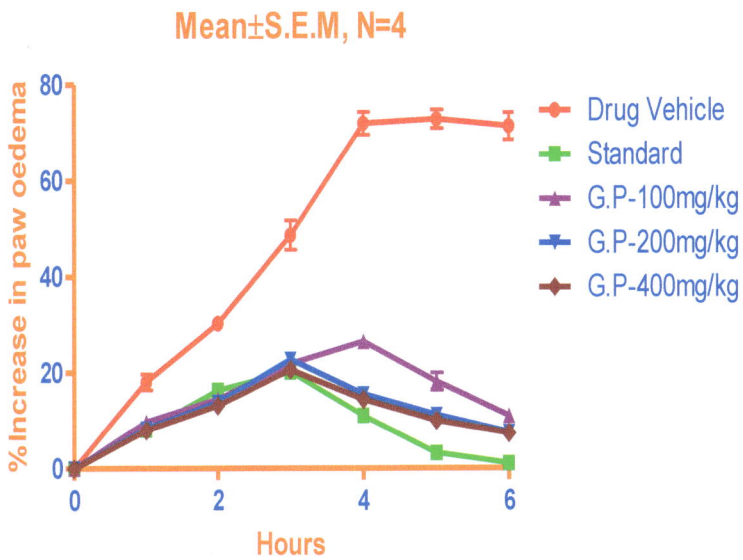

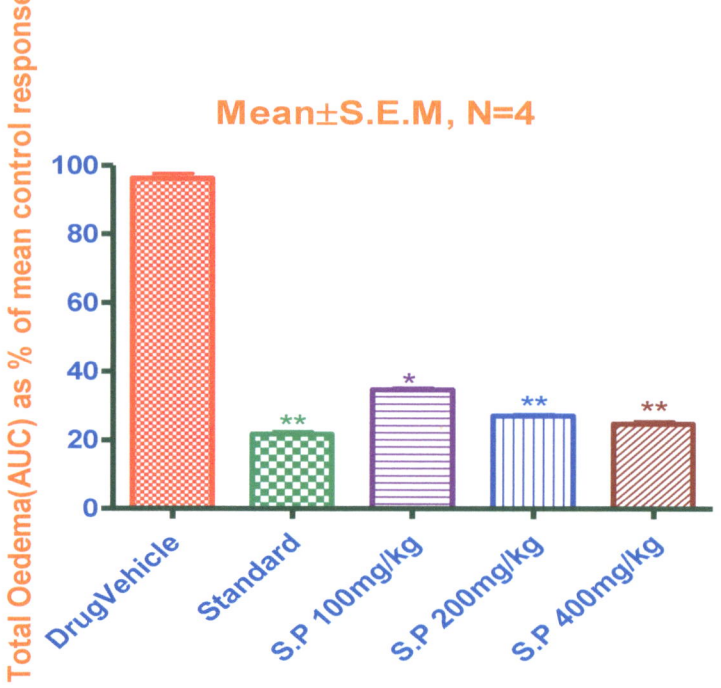

Fig-1.2.8: Effects of the crude methanol extracts of *Spondias pinnata* (stem heart wood), **100, 200 and 400 mg/kg** respectively along with Ibuprofen 50 mg/kg on A) the maximal and B) the total paw oedema in carrageenan induced rats.

Results and Discussion:

There was significant and dose dependent anti-inflammatory activity of both the ethyl acetate and methanol extracts in the acute Carrageenan-induced rat paw oedema model. Orally administered doses of 100, 200 and 400mg/kg of both ethyl acetate (64.32 ± 0.74, 69.04 ± 0.29, 72.10 ± 0.45) and methanol (63.67 ± 1.26, 68.82 ± 0.32, 71.71 ± 0.68) extracts of roots and rhizomes of *S. pinnata* produced significant reduction in paw oedema, as compared to Ibuprofen (standard)(72.41 ± 1.49) 50mg/kg. Comparatively ethyl acetate extract showed more activity than methanol extract.

Prostaglandins and bradykinins were suggested to play an important role in Carrageenan-induced oedema [28,29]. The anti-inflammatory activity which may be attributed to the presence of chemical constituents like oleanolic acid, β-sitosterol in both ethyl acetate and methanol extracts they might suppress the formation of prostaglandins and bradykinins or antagonize their action and exert its activity [31]. Further studies are needed to explore mechanism by which *S. pinnata* produces anti-inflammatory activity. We may conclude that these support the traditional use of this plant in some inflammatory and painful conditions and confirm the presence of active chemical compounds related to the activities.

1.2.22 CONCLUSION OF ANTI-INFLAMMATORY ACTIVITY:

In the current investigation ethyl acetate and methanol extracts derived from *Antigonon leptopus*, *Glycosmis pentaphylla* and *Spondias pinnata* demonstrated anti-inflammatory activity which is due to the reduction in the levels of various pro-inflammatory mediators (histamine, serotonin, prostanoids, leukotriene B_4) and free radicals (DPPH, $\cdot OH$, $\cdot O_2$, $LOO\cdot$). It indicate that all the extracts elicit different mechanisms of action against inflammation.

Carrageenan induced paw odema in acute inflammation attribute biphasic episodes incorporate of different inflammatory chemical mediators. The formation of edema is due to the release of histamine, serotonin and bradykinin after between first and second hours of carrageenan injection followed by synthesis of prostaglandin, protease and lysosomes up to sixth hour around the damage tissues.

The significant inhibitory activities of 3 plant extracts of ALEE, ALME, GPEE, GPME, SPEE and SPME on carrageenan induced paw odema at first hour indicates that the ALEE, ALME, GPEE, GPME, SPEE and SPME involve the inhibition of histamine and serotonin release. Since these plants **Antigonon leptopus, Glycosmis pentaphylla and Spondias pinnata** showed significant inhibitory effect up to fifth hour, this suggests that it may inhibit the synthesis or release of prostaglandin, protease and lysosomes. Thus all plant extracts may have an advantage over ibuprofen in suppressing inflammation.

However, the actual molecule responsible for this activity needs to be identified and pursued for future studies.

1.2.23 REFERENCES:

1) Stedman's Medical Dictionary, (1990) Twenty-fifth Edition, Williams & Wilkins.150-186.

2) Disturbance of function (functio laesa): the legendary fifth cardinal sign of inflammation, added by Galen to the four cardinal signs of Celsus. (1971) *Bull N Y Acad Med*. 47(3): 303–322.

3) *Coussens LM, Werb Z (2002). "Inflammation and cancer"*. Nature *420 (6917): 860–7.*

4) *Mohamed-Ali V et al (2001). J Clin Endocrinol Metab 86 (12): 5864–9.*

5) *Serhan CN, Savill J (2005). Nat. Immunol. 6 (12): 1191–7.*

6) The Useful Medicinal Plants of India *(ICMR),* (1978) New Delhi, 65.

7) The Medicinal Plants of India *(ICMR),* (1978) New Delhi, 165.

8) The Wealth of India "A Dictonary of India Raw Materials and Industruial Products" *(CSIR),* (1952), New Delhi, 3 and 4.

9) Handa, S.S., Kapoor V.K., Sharma, A.K., (1989), *"Text book of Phar- macognosy"* 2nd *Edn*, 57(5), 307.

10) Winter, C.A., Risely, E.A., Nuss, G.W.(1962), *Proc. Soc. Exp.Biol.Med.*,3,554.

11) Chattopadhyaya, R.N., Chattopadhyaya, R.R., Roy, S., Mitra, S.K., (1996), *Sci. Tro. Med.* 95, 45.

12) Chang, A.Y., Eydar, B.M. and Gilchrist, B.J., (1983) *Diabetes*, 32, 839-845.

13) Nasrin, M., Alireza, M.N. and Alireza, G., (2005). *Int. Immuno Pharmacol.*, Vol. 5(12), 1723-1730.

14) Iracema, E., Indira, R.S., Marcelo, R., Luis, G.V.C., Lourivaldo, S.S., Jayme, A.A.S., Fabio, F.P., Leonardo, M.L., Jose, M.S., Jairo, K.B.and Jose, C.T.C., (2005). *J. Ethnopharmacol.*, Vol. 101 (1-3), 191-196.

15) Victor, B., Owoyele, Y.Y., Oloriegbe, E.A.B. and Ayodels, O.S., (2005). *J. Ethnopharmacol.*, Vol. 99 (1), 153-156.

16) Gupta, M., Mazumder, U.K., Sambath, K.R., Gomathi, P., Rajeshwar, Y., Kakoti, B.B.and Tamil Selvan, V., (2005). *J. Ethnopharmacol.*, Vol. 98 (3), 267-273.

17) Perez-Garcia, F., Marin, E., Parella, T., Adzet, T. and Canigueral, S., (2005). *Phytomed.* Vol.12(4), 278-284.

18) Isabel, D., Jose, A.M., Ricardo, R. and Leeisa, F.M., (2004)., *Europ. J. Pharmacol.*, Vol.488 (1-3), 225-230.

19) Moura, A.C.A., Silva, E.L.F., Fraga, M.C.A., Wanderley, A.G, Afiatpour, P. and Maia,

M.B.S., (2005). *Phytomed.*, Vol. 12 (1-2),138-142.

20) Eun Mi, C. and Jae Kwan H., (2003)., *J. Ethnopharmacol.*Vol. 89 (1), 171-175.

21) Lilly, G., Yogendra, P., Richa, S., Dev, K., Sudipta, C., Mohinder, KC, Parul, B., Ravi, K. and Ramesh, C.S., (2005). *Int. Immunopharmacol.*, Vol. 5(12), 1675-1684.

22) Winter CA, Risely EA, Nuss GW. (1962) *Proc. Soc. Expt. Biol. Med.* 111; 544-547.

23) Wilhelmi G, Domenjoz R. (1951) *Arzneim-Forsch.* (Drug Res.) 1: 151-154.

24) Newbould BB. (1969) *Br. J. Pharmacol.* 35: 487-497.

25) Schayer RW, Reilley MA. (1968) *Amer. J. Physiol.* 215: 472-476.

26) Battu GR, Zeitlin IJ and Gray Al., (2000) *Br. J. Pharmacol.*, 133: 199.

27) Randall LO, Selitto JJ. (1957) *Arch. Int. Pharmacodyn.* 109: 409-419.

28) Fan, A.Y., Lao, L., Zhang, W.Y. and Berman, B.M., (2005)., *J. Ethnopharmacol.*, Vol.101 (1-3), 104-109.

29) R. Vinegar, M. Scheriber and R. Hugo, (1969) *J. Pharmacol. Exp. Ther.*, 166, 96.

30) A. Dray and M. Perkin., (1993) *Trends Neurosci.*, 16, 99.

31) K. Hemalatha and D. Satyanarayana; (2007). *Ind J. Nat. Prod.*, 23(3), 17.

32) Borgi W, Ghedira K; (2007) *Fitoterapia.* 78: 16-19.

33) Roch-Arveiller M, Giroud JP. (1979) *Pathol Biol.* 27:615.

34) Vinegar R, Traux JF, Selph JL. (1976) *Fedproc.* 35:2447.

www.ingramcontent.com/pod-product-compliance
Lightning Source LLC
Chambersburg PA
CBHW041108180526
45172CB00001B/160